NOT EXTINCT

The Przewalski's Horse Returns from Extinct in the Wild

DARCY PATTISON

NOT EXTINCT:
The Przewalski's Horse Returns from Extinct in the Wild
Copyright © 2025 by Darcy Pattison

Mims House Books
1309 Broadway
Little Rock, AR 72202
MimsHouseBooks.com

Publisher's Cataloging-In-Publication Data

Names: Pattison, Darcy, author.
Title: Not extinct : The Przewalski's horse returns from extinct in the wild / written by Darcy Pattison.
Description: Little Rock, AR: Mims House, 2025. | Summary: In 1959, the international community came together to discuss the plight of the Przewalski's horse, or P-horse. With only 12 founding horses, scientists worked together to save the P-horse from extinction.
Identifiers: LCCN: 2025902126| ISBN: 9781629442648 (hardcover) | 9781629442655 (paperback) | 9781629442662 (ebook) | 9781629443034 (audio)
Subjects: LCSH Przewalski's horse–Juvenile literature. | Endangered species–Juvenile literature. | BISAC JUVENILE NONFICTION / Animals / Horses | JUVENILE NONFICTION / Science & Nature / Environmental Conservation & Protection | JUVENILE NONFICTION / Science & Nature / Environmental Science & Ecosystems | JUVENILE NONFICTION / Science & Nature / Zoology
Classification: LCC SF363 .P38 2025 | DDC 599.72/5–dc23

ACKNOWLEDGMENTS
Thanks to Filip Mašek, Miroslav Bobek, and Barbora Dobiášová, Prague Zoo, Prague, Czechia, for information, photos, and encouragement. Thanks to Amanda Faliano and Renee Pfalzer, Denver Zoo, Denver, CO, for information and a look behind the scenes.

When a species hasn't been seen in the wild for a long time, it may be declared extinct in the wild. These animals are lost, only to be seen in zoos—if you're lucky. It's only one step away from extinction. Almost gone. Soon they'll disappear.

That's true for many species, but it's not true for…

… **the Przewalski's horse, nicknamed the P-horse! For over 65 years, scientists have fought for the return of these wild horses.**

NAME

The Przewalski's (przhě-VAWL-ski's) horse is also known as the Mongolian wild horse, Asian wild horse, the takhi (Mongolian name meaning spirit), or P-horse, for short.

A DIFFERENT SPECIES

P-horses (*Equus przewalskii*) are a different species than domestic horses (*Equus caballus*). On the next page are some similarities and differences.

P-HORSE VS. DOMESTIC HORSE

TRAITS	P-HORSE	DOMESTIC HORSE
Genus	Equus	Equus
Vocalizations	Neigh, whinny, nicker, snort, cough, grunt, snore, and squeal. They do not bray like zebras, Somali wild asses, and donkeys.	Neigh, whinny, nicker, snort, cough, grunt, snore, and squeal. They do not bray like zebras, Somali wild asses, and donkeys.
Speed	About 40 mph (65 kph), faster for short distances.	About 40 mph (65 kph), faster for short distances.
Size	Shorter and sturdier. 48–56 in (122–142 cm); 600 lbs (300 kg).	Taller and heavier. 56-72 in (142–183 cm); 900–1.200 lbs (400–550 kg).
Color	Only dun color—brown with black stockings, tail, and mane.	Many colors—browns, blacks, whites, and mixed colors.
Mane	Short, upright. No forelock (bangs).	Long, with forelock (bangs).
Wild or Not	Wild, never domesticated. It cannot be handled, mounted, or haltered.	Domesticated. Can be ridden and is often used to help humans do work. Some have gone feral.
Original Range	Native to Asian steppes.	Originated from today's Kazakhstan.
Social Structure	A herd has 1 stallion and 3-10 mares. Young stallions form bachelor groups.	

In the 1950s, P-horses were almost extinct in the wild. But international scientists agreed: the wild horses must return.

Czechoslovakia held the First International Symposium about the Przewalski's horse in 1959 at Prague Zoo.

DISCOVERED BY THE WESTERN WORLD

The Mongolian wild horse, or the takhi, was known throughout Asia for thousands of years. But in 1878, Colonel Nikolay Przewalski, a Russian geographer and explorer, brought the horse to the attention of the Western world when he returned from his journeys with the skull and hide of a horse. The newly identified species was named after him. Between 1897 and 1902, German animal dealer Carl Hagenbeck's expeditions in Mongolia captured 52 foals, which eventually lived in European zoos. During World War I, the P-horse population plunged with only 31 P-horses by 1945. Only Prague Zoo, in Prague, Czechoslovakia, and the Hellabrunn Zoo, in Munich, Germany had breeding herds.

THE FIRST STUDBOOK

Erna Mohr, a zoologist in Hamburg, Germany, established the first P-horse studbook, which listed every known living individual, its parents, where they were born, where they lived, and other information. It's like a family tree for each horse—their genealogy. At the 1959 International Symposium, Dr. Jiri Volf of Prague Zoo took over the job of studbook, which he kept for over 30 years.

Dr. Jiri Volf

To save the horses from extinction, zoologists started with a small goal: search zoos for P-horses that were the right age and physical condition to breed.

The problem? The surviving horses came from 12 original horses, and 1 final wild mare. Was it enough?

P-HORSES IN ZOOS

In 1932, the stallion Ali and the mare Minka moved from the Hellabrunn Zoo to Prague Zoo. On March 21, 1933, the foal Heluš (shown as an adult on the next page) was born. She was the first P-horse born at the Prague Zoo. She lived until 1962, mothering many foals and helping Prague Zoo become a leader in breeding the P-horse in captivity.

Studbook for Heluš.

Meanwhile, expeditions still searched for wild herds. Nothing.

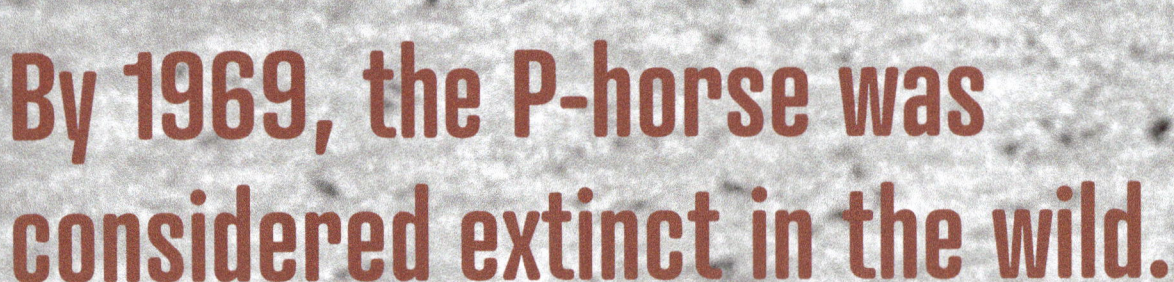

Last known photo of the P-horse in the wild.

By 1969, the P-horse was considered extinct in the wild.

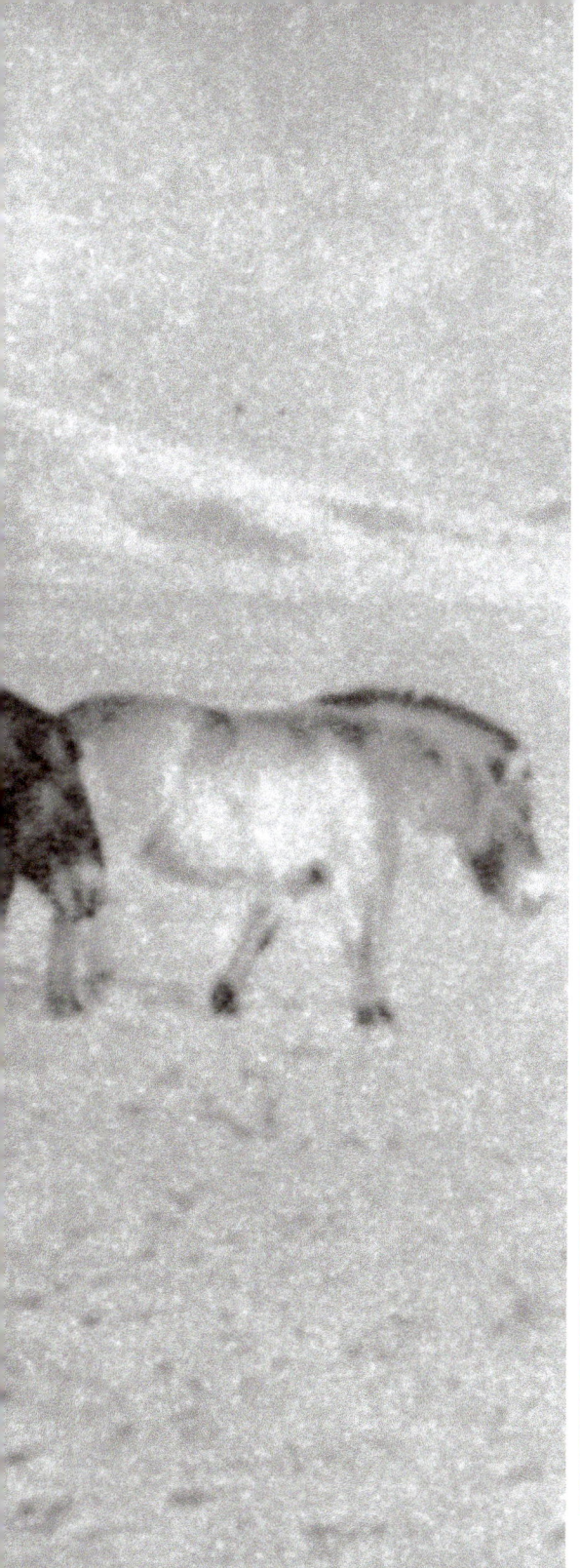

WHY DID THE P-HORSE GO EXTINCT IN THE WILD?

As human populations grew, the P-horses lost habitat. They also had to compete with domestic livestock. P-horses and domestic horses can mate and have babies. Their babies, a combination of both breeds, can also have babies. Soon, there were few purebred P-horses. Finally, in some Asian countries, it's common to eat horse meat, so they were overhunted.

IUCN RED LIST

The International Union for Conservation of Nature (https://www.iucnredlist.org) evaluates species and assigns them a category based on their risk for extinction. Starting in 1964, the IUCN Red List listed the most endangered species. In 1996, the IUCN Red List evaluated the P-horse and declared it was extinct in the wild.

THE RED LIST CATEGORIES

Extinct	No individuals survive.
Extinct in the Wild	The only indiiduals left are in captivity
Critically Endangered	Extremely high risk of extinction
Endangered	Very high risk of extinction
Vulnerable	High risk of extinction
Near Threatened	Close to being endangered in the near future
Least Concern	Unlikely to become extinct in the near future.
Data Deficient	Not enough information to categorize.
Not Evaluated	Has not been assessed.

For decades, scientists concentrated on raising P-horses in zoos or wildlife reserves. Foal by foal, the herds grew.

On April 24, 1966, a P-horse foal was born at London Zoo, the first in 36 years. The foal was named Alpha. The mother, Lunar, was received from Prague Zoo in November 1963. The father, Tzar, was born in a British location in 1957 and moved to London Zoo in 1963. International cooperation was important for the conservation efforts for the P-horse's survival.

Starting in 1985 in China and 1992 in Mongolia, scientists bred Przewalski's horses brought from Germany, the UK, and the US. The first foal was born in 1988. But these horses were kept in reserves and were provided food and water in the harsh winters. That meant the P-horses were still extinct in the wild.

Finally, after 40 years, 500 horses lived in captivity. Scientists dared to hope and dream: could they send the P-horses back to the wild?

Scientists wanted to return the P-horses to their original lands, the steppes of Asia. After much study, they decided on sites in Mongolia and China.

In 1992, after decades of waiting, a small herd of P-horses in Mongolia's Great Gobi B Reserve galloped past their last fence and onto the steppes—

the wild horses were home.

Takhin Tal, Great Gobi Reserve, Mongolia. Mares Spes, Yanja, and Helmi.

The P-horses that arrived in Mongolia in 1992 were the 12th generation of horses raised in captivity. They lived in fenced enclosures for about a year before being released to the wild.

TWO STAGES: RELEASING TO THE WILD

Returning the P-horse to the wild is a two-stage process. Upon arrival in the new environment, the P-horses are held in fenced enclosures. They must get used to the climate and weather in their new home. Here, they are fed, watered, and watched over carefully by the staff. After one to two years, the P-horses are then released into the wild.

Sending the P-horses to the wild was just the start.

To succeed, the scientists supported, studied, and observed the wild populations. They solved problems with the horses' health, supported local staff, and hoped the horses could avoid the wolves.

SEPARATION FROM DOMESTIC HORSES

The P-horse herds must be kept separate from domestic horses. If a P-horse and a domestic horse mate, their colts are a hybrid species, not a P-horse or a domestic horse. This makes it harder to save the P-horse species.

TICKS & PARASITES

One danger to the wild P-horses is diseases and parasites transmitted by domestic horses: bacteria, tapeworms, flukes, and ticks.

STAFF SUPPORT

To support the staff of the reserves, organizations like Prague Zoo donated off-road vehicles and GPS naviation, and helped build guard houses, garages, and hay barns. Without such help, the project would not have been as successful.

WOLVES

Wolves are a natural predator of P-horses. They attack colts, and up to 30% of a herd can be lost.

SOCIAL STRUCTURE

The P-horses live in social groups. A stallion will lead a herd, or harem, that includes up to 15 mares. When young stallions are two years old, the head stallion will drive them away from the group. They either join a bachelor group or try to start their own herd by stealing mares from a weaker stallion. The head stallion is responsible for the safety of his clan. He watches over the herd and alerts them when dangers arise.

STALLION FIGHTS

Stallions will sometimes fight for control. They push, bite, strike with hooves, and kick. Only one stallion is in charge of the mares at any point. Younger or stronger stallions must fight for the control of the herd.

But then, in 2009 and 2010, two-thirds of the P-horses in Mongolia's Great Gobi reserve died from a dzud, the Mongolian word for extreme winter. The scientists grieved and worried. Could they save the P-horse from extinction?

Or was the wild just too wild?

SURVIVING THE WINTER

P-horses live in cold, windswept prairies, or steppes. As herbivores, they eat grass and other plants. They thrive in cold weather by growing thick winter coats and learning to dig down for grass to eat. The shaggy winter coats that protect them from the cold are worn for eight months of the year.

But some winters are too harsh. In 2009 and 2010, summer droughts meant there was little grass available for grazing. Then bitterly cold winter temperatures made it very hard for animals to find food. Mongolian herders lost up to 53% of their domestic animals. Across the country, 8–10 million animals died from the cold. The P-horses were also affected, and the losses among this endangered species were devastating.

The P-horses are loaded onto cargo planes.

The scientists realized they were relying too much on only a few wild locations. Again, scientists searched the Asian steppes for grassy areas with plenty of water—places where the P-horses could run wild and free. In 2024, seven horses raced across the Kazakhstan plains, the golden steppe, a new home.

Prague Zoo started the Return of the Wild Horses project to return the P-horse to the Asian steppes. With help from Czech Army airplanes, they transported P-horses from Europe to Mongolia for nine years. Prague Zoo also led the way in choosing the new Kazakhstan location for the P-horse.

Transporting P-horses in Kazakhstan.

TRANSPORTING

To transport the European P-horse to Asia is difficult and dangerous. The P-horses are loaded into specially designed boxes that force the P-horses to stand upright. Scientists have learned that if a P-horse sits or lies down while traveling, their blood flow can be interrupted, and the horse may die.

Below: Miroslav Bobek, Director of Prague Zoo, cheers as the first P-horse steps onto the golden steppe of Kazakhstan.

A P-horse travel box.

Across the world, over 3,000 P-horses now live in about 120 zoos and wild reserves.

After 65 years of hard work, P-horses have moved from extinct in the wild, to critically dndangered, to only endangered.

What a triumph!

Left: These are the first P-horses in central Kazakhstan after hundreds of years. Pictured here are three horses sent from Czechia: the mare Ypsilonka (left), the mare Zeta II (middle), ; and the stallion Zorro (right).

Are the scientists done yet? No.

The P-horse is still considered an endangered species. The fight for this amazing animal continues.

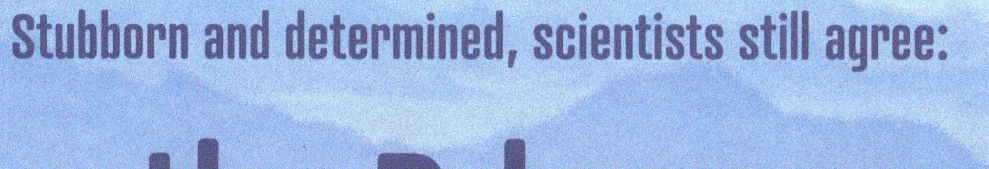

Stubborn and determined, scientists still agree:

the P-horses must return to the wild.

PHOTO CREDITS

Cover: Miroslav Bobek, Koně Přzevalshénko v Dolnim Dobřejově [Przewalski's horses in Dolni Dobřejov], © 2025 Prague Zoo.

Pages 2-3: Kate Jenks, Przewalski's Horses in Hustai Park, Mongolia, © 2016.

Pages 4-5: (top) Bay Horse, OlgaIT / iStock Photo; (bottom) Przewalski's Horse, MariaItina / Alamy Stock Photo.

Page 6: First International Symposium About the Przewalski's Horse, courtesy of Prague Zoo.

Page 7: (top) Nikolay Przhevalsky, public domain, downloaded from Wikipedia Commons; (bottom) Dr. Jiri Volf, courtesy of Prague Zoo.

Page 8: Ali and Minka, courtesy of Prague Zoo.

Page 9: (top) Heluš's Studbook, courtesy of Prague Zoo; (bottom) Heluš, courtesy of Prague Zoo.

Pages 10-11: Last Known Photo of the Przewalski's Horse in the Wild, courtesy of Prague Zoo.

Page 12: May 05, 1966 - The First Mongolian Wild Horse Foal to Be Born at the London Zoo for 36 Years, KEYSTONE Pictures USA / Alamy Stock Photo.

Page 13: Per Hamernik. Hřebeček Koně Přzevalshénko x Matkou Harou [Przewalski's horse colt x mother Harou], January 2005, © Prague Zoo.

Page 14: Miroslav Bobek, Mare and foal, December 2024, © Prague Zoo.

Page 15: Closeup of Przewalski's Horse, Xinhua / Alamy Stock Photo.

Pages 16-17: Przewalski's Horses Galloping Across the Steppes, Terry Allen / Alamy Stock Photo.

Page 18: Jaroslav Šimek, Návrat Divokých Koni [Return of the Wild Horses], © 2019 Prague Zoo.

Page 19: (top) Petr Josek, Releasing Przewalski's Horses, © 2014 Prague Zoo; (bottom) Miroslav Bobek, Two Przewalski's Horses Grazing, © 2017 Prague Zoo.

Pages 20-21: Václav Šilha, Four Przewalski's Horses, © 2017 Prague Zoo.

Page 21: Uwe Aranas, Dalian, Liaoning, China: A Wolf (Canis lupus) in Dalian Forest Zoo © CEphoto, Uwe Aranas, CC BY-SA 3.0, https://commons.wikimedia.org/w/index.php?curid=42775024.

Page 22: Head of Przewalski's Horse, Tierfotoagentur / Alamy Stock Photo.

Page 23: Przewalski's Horses Fighting, Juniors Bildachiv GmbH / Alamy Stock Photo.

Pages 24-25: Przewalski's Horse Herd, Juniors Bildachiv GmbH / Alamy Stock Photo.

Page 26: Miroslav Bobek, Loading Przewalski's Horses onto Airplane, © 2024 Prague Zoo.

Page 27: (top) Miroslav Bobek, Truck Loaded with Przewalski's Horse Driving Across Kazakhstan Steppes, © 2024 Prague Zoo; (bottom) Václav Šilha, Releasing Przewalski's Horses in Kazakhstan, © 2024 Prague Zoo; (right) Miroslav Bobek, Przewalski's Horse Loading into Travel Container, © 2024 Prague Zoo.

Pages 28-29: Miroslav Bobek, Herd Racing, © 2018 Prague Zoo.

Page 28: Miroslav Bobek, Three Przewalski's Horses, © 2024 Prague Zoo.

Page 29: Miroslav Bobek, Przewalski's Horses in Kazakhstan, © 2024 Prague Zoo.

Pages 30-31: Miroslav Bobek, Návrat Divokých Koni [Return of the Wild Horses], © 2019 Prague Zoo.

Page 32: Przewalski's Horse – Silhouette of Head, Imago / Alamy Stock Photo.

FOR MORE ON THE PRZEWALSKI'S HORSE - VIDEOS

"Przewalski's Horse: Subspecies Returned to Wilds After Near Extinction," Al Jazeera English, June 13, 2024. https://youtu.be/1fMz9nfwLxs.

SOURCES - INTERVIEWS

On-site interview by the author at Prague Zoo, Prague, Czechia, with Filip Mašek, Spokesperson, September 9, 2024.

On-site interview by the author at Denver Zoo Conservation Alliance, Denver, CO, with Reese Pfalzer, Assistant Curator of Hoofstock, September 27, 2024.

Email interview with Amanda Faliano, Denver Zoo Conservation Alliance, Asian Wild Horse Studbook Keeper, Asian Wild Horse Vice Coordinator. Several emails, October 2024.

Email interview with Barbora Dobiášová, Curator of Ungulates, EEP coordinator for Przewalski's horse, Prague Zoo, Prague, Czechia. Several emails, October 2024.

SOURCES - BOOKS

Boyd, Lee, and Katherine A. Houpt. Przewalski's Horse: The History and Biology of an Endangered Species. Albany, NY: State University of New York Press, 1994.

www.ingramcontent.com/pod-product-compliance
Lightning Source LLC
Chambersburg PA
CBHW040225040426
42333CB00054B/3451